HANNACHI Abdelhakim
ZOUITEN Khaoula
KHORIEF Maissa

Modeling tillering in a durum wheat crop Triticum durum L

HANNACHI Abdelhakim
ZOUITEN Khaoula
KHORIEF Maissa

Modeling tillering in a durum wheat crop Triticum durum L

Imprint
Any brand names and product names mentioned in this book are subject to trademark, brand or patent protection and are trademarks or registered trademarks of their respective holders. The use of brand names, product names, common names, trade names, product descriptions etc. even without a particular marking in this work is in no way to be construed to mean that such names may be regarded as unrestricted in respect of trademark and brand protection legislation and could thus be used by anyone.

Cover image: www.ingimage.com

This book is a translation from the original published under ISBN 978-620-6-71115-5.

Publisher:
Sciencia Scripts
is a trademark of
Dodo Books Indian Ocean Ltd. and OmniScriptum S.R.L publishing group

120 High Road, East Finchley, London, N2 9ED, United Kingdom
Str. Armeneasca 28/1, office 1, Chisinau MD-2012, Republic of Moldova, Europe
Printed at: see last page
ISBN: 978-620-8-16268-9

Copyright © HANNACHI Abdelhakim, ZOUITEN Khaoula, KHORIEF Maissa
Copyright © 2024 Dodo Books Indian Ocean Ltd. and OmniScriptum S.R.L publishing group

Contents

Resume

Scientists have carried out research and experiments over the years in an attempt to improve the production of durum wheat (*Triticum durum* L.) and achieve food self-sufficiency, considering it to be a staple food.

In our experiment, we used four varieties of durum wheat (*Triticum durum* L.) most commonly grown in Algeria (Simeto, Oued el Bared, Amare 06, Antalis). We carried out this experiment in the garden of the University of 20 August 1955 Skikda, out of soil and under natural conditions. During our analysis, we carried out a complete study to model the components of durum wheat (*Triticum durum* L.), using nitrogen as the main element to compare the results before and after the use of this substance.

The results showed that the varieties responded linearly to this element and gave high results, particularly for the Simeto variety, where we recorded a multiplication of the yield components and therefore an increase in yield. We can therefore conclude that there was an increase in the number of yield components at the heading stage.

Keywords: Modelling, Tallage, Fertilisation azotee, Ble dur, Rendement.

Introduction

Nowadays, cereals in general, and wheat (durum) in particular, are the main staple diet for Algerian consumers (Benbelkacem, 2013). It plays a social, economic and political role in most countries in the world (Ammar, 2015 in Bentouati, Safsaf, 2019).

Cereals are the main crop grown in Algeria. It is grown mainly in arid and semi-arid areas, on an annual area of around 3.6 million hectares (ONFAA, 2016).

Hard wheat breeders have made unique contributions and excellent progress in increasing production over the past decades, mainly in less developed countries. However, there are still many challenges ahead to make food more accessible than ever in a sustainable manner and to meet the needs of a growing population (IWGSC, 2019).

Productivity and resource acquisition (sustainable development) are receiving the most attention at the moment and it is very important and necessary to work on the development of agricultural skills for forecasting and decision-making in agriculture and forestry and the dynamic development of reliable mathematical models, such as crop simulation and improvement. And all this thanks to the work of engineers and researchers who represent the most effective and integral part of this development, and these results seek to reproduce the growth of a group of plants in interaction with the environment, to determine the impact of the monastery can have these factors. And in the long term, to make crop predictions (Barcziet *al.* 1997).

However, modelling entered the field of agronomy in 1968, and modelling plant growth remains a challenge, requiring collaboration and confrontation between several disciplines. Indeed, at the time of the

Designing such a model requires a combination of botanical, agronomic and ecophysiological knowledge.

Depending on the objectives being pursued, the essential elements must be extracted and summarised. Next, validation and exploitation require computer implementation of the model, and the introduction of a mathematical formalism into certain models enables their behaviour to be verified and more in-depth studies to be carried out (optimal control, sensitivity to parameters) (Brisson et *al.* 2003).

Our subject involves modelling tillering in a durum wheat crop, under the effect of nitrogen.

How can we improve tillering models in durum wheat?

J How are we going to create a mathematical model?

What analysis will we use to find the difference between the trials? What is the effect of nitrogen treatment on tillering and development of hard wheat?

Our work is divided into two distinct parts

The 1ere bibliographical review is divided into two chapters

Chapter 01: General information on hard disk

Chapter II: General information on modelling tillering in a durum wheat crop.

The 2eme part is the experimental part divided into two chapters

Chapter I: Materials and methods

Chapter II: Results and discussion

General information on durum wheat (*Triticumdurum*)

1- History and geographical distribution of durum wheat (*Triticumdurum*)

Since the birth of agriculture, hard wheat has been the staple food of mankind (Ruel, 2006). The discovery of wheat dates back to 15,000 BC in the region of the Fertile Crescent, a vast territory comprising the Jordan Valley and adjacent areas of Palestine, Jordan, Iraq and the western edge of Iran (Feldman and Sears, 1981). This corresponds to the beginning of the Dryas period, which was a climatic episode of drought and cooling, leading to the gradual end of the 'hunter-gatherer' way of life and to the domestication of certain plants - including wheat - and, via the storage of food stocks, the creation of the first village communities (Hayden, 1990; Wadley and Martin, 1993).

At first, bles evolved without human intervention, then under the selection pressure exerted by the first farmers (Henry and de Buyser, 2001). In a first phase, which corresponds to the transition period between the manual collection of wild forms in their native habitat and the appearance of the first cultivated fields, the transition from fragile ear forms to solid rachis types was decisive, as was the identification of mutants with easily threshed ears and naked grain. This period was also accompanied by other changes, such as a preference for ericaceous plants with large, non-dormant grains that germinated uniformly, and certainly a selection based on the colour of the grain, linked to religious or other practices. It is also possible that, from this stage onwards, farmers became aware of the importance of the number of spikelets per ear, but this is not certain (Bonjean, 2001).

It is generally accepted that the cultivation of durum wheat began and developed in Algeria after the Arab conquest. Most authors agree that Algerian cereal growing was dominated by durum wheat from that date until colonisation (Laumont and Eurroux, 1961).

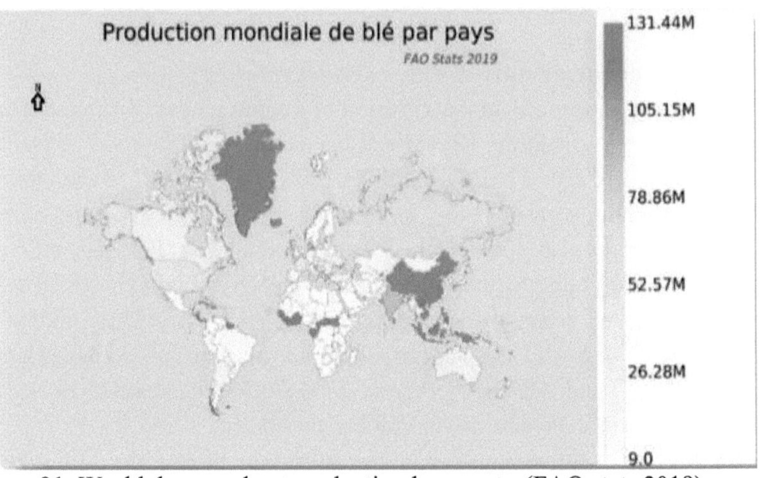

Figure 01. World durum wheat production by country(FAO stats 2019)

2- Genetic origin of durum wheat (*Triticumdurum*)

Wheat belongs to the gramineae family (Gramineae = Poaceae), which includes more than 10,000 different species (Mac Key, 2005). Several species with different ploidies are grouped together in the genus Triticum, which is a classic example of allopolyploidy, with homeologous genomes derived from hybridisation between species belonging to the same family (Levy and Feldman, 2002).

According to Feillet (2000), these species differ in their degree of ploidy (diploid bles: genome AA; tetraploid bles: genomes AA and BB; hexaploid bles: genomes AA, BB and DD) and in their number of chromosomes (14, 28 or 42). The polyploid nature of the bles genome is also thought to have contributed to their domestication success (Dubcovsky and Dvorak, 2007).

The genetic lineage of bles is complex and incompletely elucidated. It is known that the A genome comes from Triticum. Monococcum, genome B from an Aegilops (bicornis, speltoides, longissima or searsii) and genome D from Aegilops squarrosa (also known as *Triticumdurum*. Tauschii). The natural cross *Triticumdurum. Monococcum* **xAegilops** (carrying the B genome) resulted in the appearance of a wild AABB wheat *(Triticumdurum*. Turgidumssp. Dicoccoides) which then gradually evolved into *Triticumdurum*. Turgidumssp. Dicoccum and then to *Triticum. Durum* (hard wheat) (Feillet, 2000). Remains of primitive types of *Triticumdurum*. Turgidum cultivum (the starch tree, which is a coarse-grained wheat), discovered at several archaeological sites in Syria, have been dated to around 8000 BC (Brink and Belay, 2006). The cross between the species *TriticumDurum* with genomic constitution AABB and the Aegilops tauschiide with genomic constitution DD, gave rise to the species *Triticumdurum*. Aestivum with genomic constitution AABBDD (Feldman and Sears, 1981; Shewry, 2009) (Figure 01).

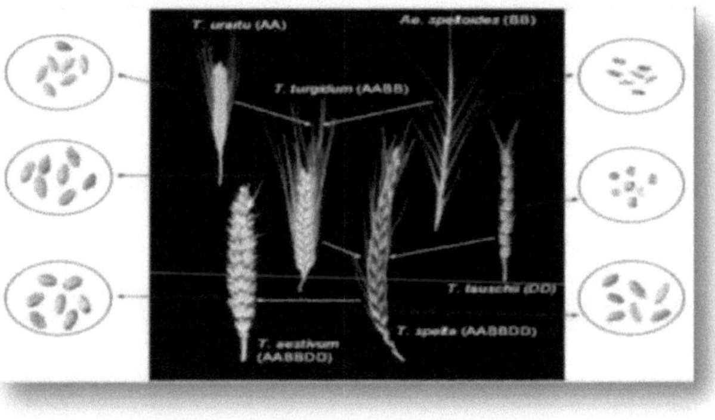

Figure 02. Phylogeny of hard wheat (Shewry, 2009)

3- Botanical classification of durum wheat (*Triticumdurum*)

Ble dur belongs to the Spermaphytes group and to the Spermaphytes group. Angiosperms, in the class of monocotyledons (Grignac, 1965; Prats, 1966). The classification proposed by Feillet (2000) is shown in table 01.

Table 01Botanical classification of hard wheat (Douaer et al.2018)

Regne	**Plantae**
Sub-regne	*Cormophyte*
Branch	*Spermaphytes*
Sub-branch	*Angiosperms*
Class	*Monocotyledone s*
Order	*Floral Commelini*
Under order	*Poales*
Family	*Graminees*
Tribe	*Triticees*
Type	*Triticum*
Species	*Triticumdurum*

4- Morphological characteristics of hard wheat

(*Triticumdurum*)4-1-Grain

The wheat kernel is ovoid in shape and has a groove running along its entire length on its ventral side (Figure 0 3). At the dorsal base of the seed is the germ, which is topped by a brush. It measures between 5 and 7 mm long, and between 2.5 and 3.5 mm thick, weighing *between 20 and 50 mg (*Surget and Barron 2005).

According to Calvel *(1983),* the colour of blue varies from red to white. In relation to the country of origin, soil, cultivation and climate (Emilie, 2007)

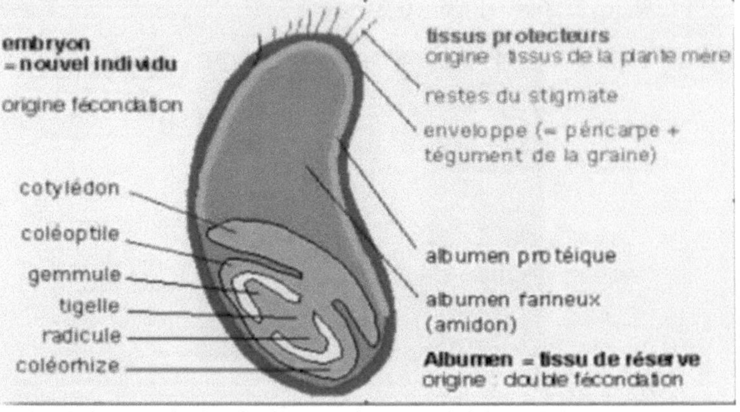

Figure 03. Schema of a hard wheat grain (Mossiniak, 2006)

4-2-Racine

The roots of hard wheat are of the poorly developed fascicle type, the root system of hard wheat is characterised by (Alismail et al., 2017)

> **A primary system (seminal roots)** this root system functions from germination to branching of the plant, i.e. till tillering (Grignac, 1965), there are 6 seminal roots (Colnenne et al, 1988), which participate in the nutrition and development of the plant (Belaid, 1987).

> **A secondary system (adventitious roots)** which form later from the nmuds at the base of the plant and constitute the permanent root system, (Clarke et al. 2002). According to Grignac et al (1965), the secondary system is a tillering root system. The roots appear lower down, almost all at the same level (tillering plateau), and form a dense clump. Each tiller gives rise to a culm and an inflorescence (Belaid, 1987).

4-3-Stem

Most hard wheat plants have a main stem and secondary stems called tillers. The stems are cylindrical culms, often hollow due to resorption of the central pith, but in hard wheat they are solid. They appear as grooved tubes, with long, numerous sap-conducting bundles. These bundles are regularly criss-crossed and contain thick-walled fibres that give the structure strength.

The culms are interrupted by nodes, which are a succession of zones from which a long leaf emerges (Soltner, 1990). The stem begins to take on its character at the start of bolting, i.e. becomes vigorous and bears 7 to 8 leaves (Alismail et al, 2017). Depending on species variation and environment, the length of the complete stem varies, but in general it ranges between 60 and 150 cm

(Mohamed, 2000). Talle production begins once the third leaf has developed, around 45 days after sowing (Moule, 1971 in Nadjem, 2012).

According to Mohamed (2000), the number of tillers in hard wheat varies from 30 to 100. This is influenced by several factors, the most important of which are variety, soil fertility, plant density and light intensity. The plant generally bears 2 to 3 tillers under favourable conditions. And tillering stops temporarily as the stem elongates.

4-4-Leaf

The leaf of the hard wheat is simple, elongated, alternate and parallel-veined (Oudjani, 2008), and consists of a base (sheath) surrounding the stem, a terminal part that aligns with the parallel veins and a pointed tip. At the point of attachment of the leaf sheath is a thin, transparent membrane (ligule) with two small lateral appendages (auricles). The main stem and each sprig bear a terminal spike inflorescence. (Cherfia, 2010). According to Casnin et *al.* 2013, the size of the leaf increases with its position on the stem, the flag leaf often being the largest. It is about 30 cm2 in size, and when mature, the wheat plant has about 1.5 to 2 m .^{2}

4-5-Flower

The flowers are numerous, small and inconspicuous (Figure 06). Located at the tip of the culms (Sadoukiet al , 2018), they are grouped into a spike inflorescence, the basic morphological unit of which is the spikelet consisting of a cluster of flowers enveloped by their glumes and enclosed in two bracts called the glumes (lower and upper) (Gate, 1995).

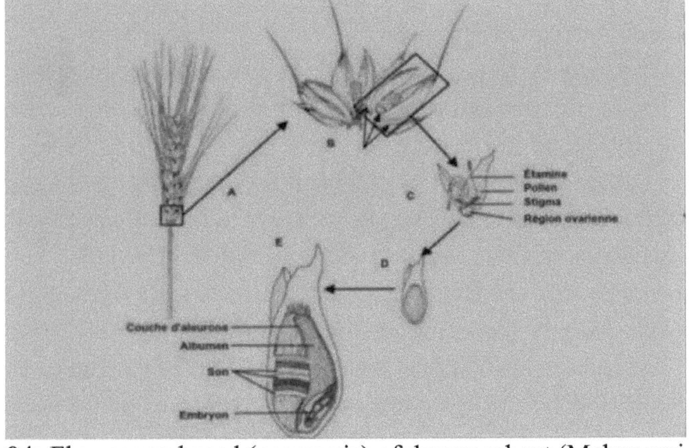

Figure 04: Flowers and seed (caryopsis) of durum wheat (Mekaoussi, 2015).

5- The biological cycle of durum wheat (*Triticumdurum*)

From seed to seed, the biological cycle of durum wheat is divided into three successive periods, each with its own phases and stages (Fig. 02).

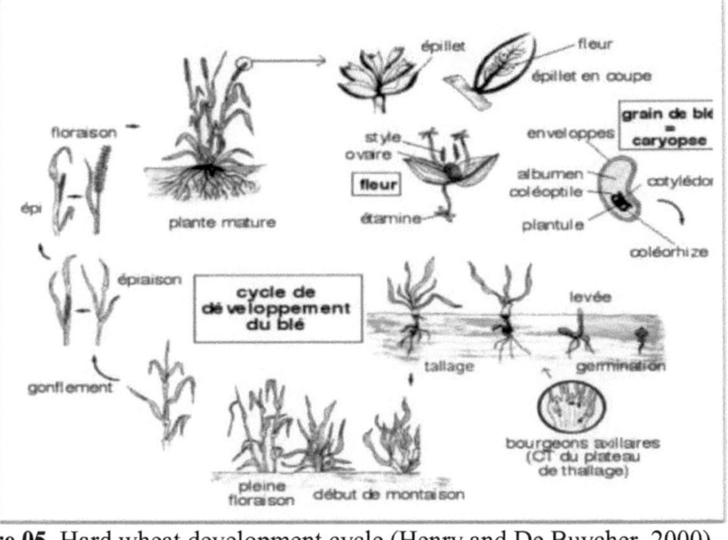

Figure 05. Hard wheat development cycle (Henry and De Buycher, 2000)

The different stages are controlled by the sum of the daily temperatures (degree-days) experienced by the plant.

The sum of the temperatures, the zero base for the blue, is calculated as follows sum degree-day = (T°C min+T°C max)/2 Only positive values should be taken into account (Hamadache, 2013).

5-1-Vegetative period

> Germination and sprouting phase

This phase corresponds to the establishment of the number of plants/m^2 . The soil is pierced by the coleoptile, which is a protective sheath for the first leaf (Hamadache, 2013).

Emergence is noted when 50% of the plants have emerged from the soil (Figure I.3). During this phase, young plants are sensitive to lack of water, which causes plant loss, and to cold, which causes heaving (Karou et *al.* 1998).

This phase begins with the fourth leaf. The first tillers are formed at the 3-leaf stage. The first primary shoot (master shoot) appears in the axil of the first leaf of the ble. The 2nd and 3rd[eme] tillers appear in the axils of the 2nd and 3rd leaves (Hamadache, 2013). Late tillering marks the end of the vegetative period and the start of the reproductive phase, which depends on the photoperiod and vernalisation, which allow the internodes to elongate (Gate, 1995).

However, (Longnecker et *al* 1993) suggest that tillering does not stop at any stage of wheat development, but is rather controlled by a number of genetic and environmental factors. The number of productive tillers depends on genotype, environment and is strongly influenced by stand density (Acevedo et *al.*, 2002).

5-2- Reproductive period

> Setting - flowering

Bolting begins when the internodes of the main stem detach from the tillering plateau (Belaid, 1987). According to Baldy (1984) bolting is the most critical phase in the development of wheat. Any hydric or thermal stress during this phase reduces the number of rising ears per unit area.

At heading, the ear emerges from the last leaf. Dehulled ears generally flower within a few days (less than 7 days) of heading.

High temperatures and drought during heading and flowering can reduce the viability of pollen and thus reduce the number of grains (Herbek and Lee, 2009).

> Flowering-maturity

The flowering-maturity period corresponds to the accumulation of carbohydrates and nitrogen in the grain (Gallais and Bannerot, 1992). This period corresponds to the formation of the 8dern component of yield, which is the weight of 1000 grains (Robert et *al.*, 1993). After flowering, the grain fills up in two stages

> By the migration of part of the stem reserves.

> Grain yield, under a rainfed cropping system and a restrictive environment, is the result of the duration and speed of filling and the capacity to transfer assimilates stored in the stem (Abbassenne et *al.*, 1997).

> High temperatures during this period stop the migration of reserves from the leaves and stem to the grain (grain scald). The grain then dries out to reach its final dry weight (Wardlow, 2002).

Grain yield, under a rainfed cropping system and a restrictive environment, is the result of the duration, filling speed and translocation capacity of the assimilates stored in the stem (Abbassenne et *al.*, 1997).

High temperatures during this period stop the migration of reserves from the leaves and stem to the grain (grain scald). The grain then dries out to its final dry weight (Wardlow, 2002).

1- Ecological requirements for hard wheat
(*Triticum durum L.*)

6-1- Water

Water is an environmental factor that influences almost all physiological reactions in plants. According to Duthil (1973) and Catell (2006), studies show that water is the main constituent of plants, accounting for 60% to 80% of their weight in fresh matter. The early grain can absorb 40 to 65% of its weight in water, but germination begins when it has absorbed around 25% (Prats et al.,

1971). The water requirements of durum wheat depend on its development cycle and the different phases that make it up (Merouche et al. 2015). Water requirements up to the end of tillering are relatively low, while a good water supply is particularly important between heading and flowering and between the milky grain and coarse grain stages (Clement, 1981). According to Merouche et al. (2015), water consumption in durum wheat is divided as follows:

J **In the epi 1cm - 2 nwuds phase:** the water supply lasts 20 to 25 days and is 60 mm high.

J **In the 2-node - flowering phase:** water consumption is around 160 mm and takes 30 to 40 days.

J **In flowering phase - milky grain**: water needs 20 to 25 days and is 140 mm.

J **In the milky grain - ripening phase**: an average of 90 mm is sufficient for a period of 15 to 20 days.

6-2- Temperature

Like all plants, hard wheat has an ecological optimum from which temperature (Papadakis, 1932), germination begins as soon as the temperature exceeds 0°C, with optimum growth temperature situated between 15 and 22° C, (OE Ondo, 2014). Temperature plays an essential role in plant life, determining growth and development (Kamli, 1985). Its action is permanent throughout the cycle. It determines nutrient uptake, photosynthetic activity, dry matter accumulation and the transition from one vegetative stage to another (Van Oosterom et al.1993 ; Mekhlouf et al., 2006).

Wheat seeds require an approximate temperature range of 80°C to 122°C to germinate. The higher the average temperature, the shorter the germination time (Lounis et al. 2017).

The demand for hard wheat in temperature: (Lounis et al, 2017):

J **Tillering phase** Starts at 2°C to 3°C and increases between 15°C and 19°C.

J **Harvest phaseThe** optimum temperature for this phase is between 16°C and 18°C.

Both phases require temperatures between 20°C and 22°C.

Note that the first step is to choose the right sowing time to avoid any damage caused by critical and early weather (Lounis et al., 2017).

6-3- Lighting

Wheat is a long-day plant. Light is considered to be a climatic parameter that plays a part in the photosynthesis phenomenon, as straw cereals are C3 plants that do not require much light (Merouche et al. ,2015).light is the source of energy that enables the plant to decompose atmospheric CO2 in order to assimilate carbon and carry out carbohydrate photosynthesis. According to Baldy (1992), light increases the ability of hard wheat to branch and increases the amount of dry matter. So, for good tillering, the wheat must be placed in optimum light conditions. A precise day length (photoperiodism) is necessary for flowering and plant development (Gouasmi and Badaoui, 2017). Often, early growth requires low light intensity (500 to 1000 lux) with a photoperiod of 12 to 16 hours of light (Boukensous et al.2014).

6-4- Soil

The soil is the support for vegetation, its larder and its water reservoir (Girard et al.2005).The soil acts through its physical, chemical and biological properties. Its composition of mineral elements, organic matter and structure play an important role in plant nutrition, and thus in determining seed yield expectancy (Olioso,

2006). Hard wheat requires well-prepared, loosened soil to a depth of 12 to 15 cm for loamy soils and 20 to 25 cm for other soils (Ouanzar, 2012).

A clayey-limestone or silty-clayey soil favours rooting in hard wheat (Merouche et al., 2015), whereas a light-textured, acidic soil, or one containing high levels of sodium, magnesium or iron, is unfavourable for hard wheat cultivation (Merouche et al., 2015).

Merouche et al (2015), a pH value between 6.5 and 7.5 seems to be favourable for wheat crops. However, durum wheat is sensitive to limescale and salinity, and salt can adversely affect germination rates, biological growth and grain production (Merouche et al., 2015).

6-5- Fertilisation

Nitrogen-phosphorus fertilisation is very important in Saharan regions with skeletal soils. Hard wheat needs the following three essential elements

6-5-1-Nitrogen (N)

It is a very important element in the development of blueberry (Viaux, 1980), allowing the multiplication and elongation of leaves and stems, and an increase in vegetative mass.

6-5-2-Phosphorus (P)

Theoretical phosphorus requirements are estimated at around 120 kg P2O5/ha (Balaid, 1987 in Ouanzar, 2012).

6-5-3-Potassium (K)

Cereals may require more potassium than the amount contained at harvest: 30 to 50 kg K/ha (Balaid, 1987 in Ouanzar, 2012).

6-6-Varnishing

It consists of a period of low temperatures, which cereals need for floral initiation in autumn (Boulal et al. 2007).

2- Importance and production of durum wheat (Triticumdurum) in the world and in Algeria

Durum wheat is a secondary cereal crop worldwide. It is produced mainly in the Mediterranean basin (southern Europe, Middle East, North Africa) and in North America (central Canada and northern USA), where a quarter of the world's durum wheat is produced (Clerget. 2011). Wheat is a cereal with very high economic stakes. In volume harvested, with an estimated 2518.8Men 2013/2014.

> **In Algeria**, wheat is the country's leading cereal crop. It is grown on more than a million hectares a year. Cereal production for the 2012/2013 season totalled 49.1 million quintals at national level, down 9,000,000 quintals on the previous season. Drought hit cereal-growing areas in the east of the country in 2013. This resulted in a 25% increase in the quantities of cereals imported by Algeria. Nevertheless, this increase in terms of quantities did not affect the import factor, which fell slightly by 0.6% compared with the previous year.

Imports totalled 3.16 billion dollars in 2013, compared with 3.18 billion dollars for the same period in 2012, down 0.62%.Despite tripling its production since the country's independence in 1962, Algeria remains one of the world's biggest cereal importers (Amarn, 2014).

3- Le Tallage

In botany and in the Poaceae (Gramineae), tillering distinguishes between lower nodes that develop adventitious roots and adventitious buds that lead to the formation of a tuft. This is the formation of lateral shoots, which originate from basal, underground nodes, generally those located closest to the surface of the earth (Michel and Jean-Louis, 2011). Mode of development of certain gramineae (most straw cereals), which consists of the formation of a tillering plateau followed by the emission of tillers. Finally, the number of tillers emitted per plant characterises herbaceous tillering. This will depend mainly on: the species, the variety used, the climate (temperatures) of the year or the region, the plant's nutrition, and the depth of sowing.

8-1- The origin of tillers

According to Moule (1971), tillering is characterised by the entry into growth of different buds in the axil of each of the first leaves: it is therefore a simple branching process. The first tillers (T1) generally appear in the axil of the first leaf when the plant is at the "4-leaf" stage. This tillers consists of a pre-leaf surrounding the first functional leaf of the tillers, which itself covers the others, followed by the 2nd, 3rd,[eme] and 4[eme] leaf tillers (Figure 09), from the buds that originate in the axils of the corresponding leaves. These tillers are known as primary tillers.

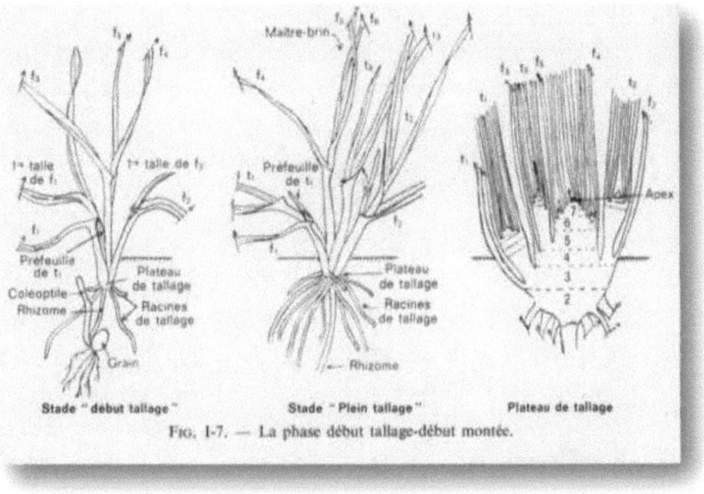

Fᴵᴳ. I-7. — La phase début tallage-début montée.

Figure 06. The tillering plateau, the area around the crown where the tillers are emitted (Agric, 1977).

Each primary tillers will give rise to secondary tillers which in turn give rise to tallestertiaries.

The ability to produce a greater or lesser number of secondary and tertiary tillers is both a specific and a varietal characteristic.

8-2-The formation of the tillering plateau

According to Ducreux (2002), in cereal plants, branching occurs at the point of contact between the root and the stem, and there can be up to 30 branches, depending on the type of plant (Evan, 1975). After the emergence of the third leaf, a phenomenon known as "Pre-Tallage" occurs, whereby the second phalanx carrying the terminal bud is elongated to the point where it can be seen on the stem.

Inside the Coleoptile (Jonard, 1951), it stops 2 cm below the surface whatever the depth of the plant. New roots appear at the fourth leaf stage with the emergence of the first tillers at the base of the branch.

According to Michele et al (2006), after germination begins, node 2 (E2) lengthens considerably, reaching two cm from the surface. The following internodes (E3, E4) remain short. The axillary buds formed in the leaf axils (F1, F2) will develop to give new leafy stems, the tillers (T1, T2), thus forming the tillering tray.

8-3-Tallage bed architecture

The tillers are identified at the leaf axil, or prophyll, or at the Coleoptile that has emerged from it (emergent) (Bos et al. 1998). The tillers that grow from the buds in the tillering plate are called primary tillers, the initial tillering is determined from the axil of the main buds of the first leaves of the main stem and called (T1, T2, T3   ), the secondary tillers that grow from the axil of the leaflet of T1, T2 and called T1.1, T2.2, T3.3, etc., while the tillers produced from the prophyll are called T1.0 T2.0 T3.0 ...,(Moeller et al., 2014) (Figure 10).

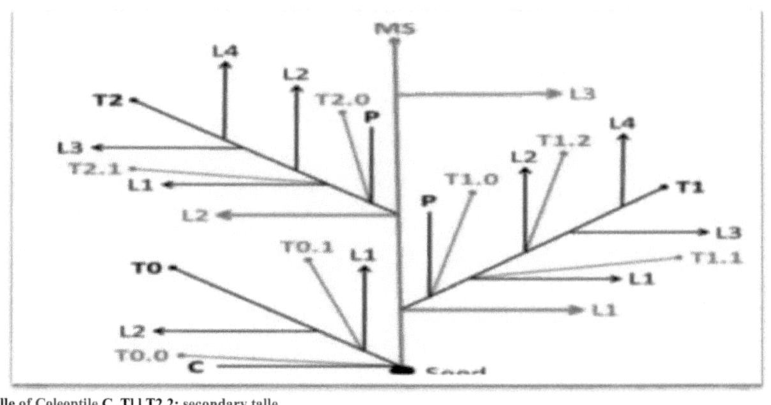

TOtalle of Coleoptile C. Tl.l T2.2: secondary talle
LI, L2. L3 leaf. T1.0 T2.0: P. prophyll heel.
T1 T2 T3: talle premiere. MS: main stem. T: talle. L: leaves. C: *Coleoptile,*

Figure 07. The formation or architecture on the stemMoller et *al,* 2014inZeddig, 2019

4- Factors that influence and promote tillering

According to Moule (1971) the number of potential tillers depends on the following factors:

J The choice of the variety of the type of hard wheat the varieties differ in their tillering capacity and their speed of development.

Early sowing increases the number of tillers per plant.

Seed rate planted a high rate produces more main stems and fewer tillers.

Soil/seedbed conditioning Poor seedbed quality and compaction slow initial till development.

Pest control against slugs, for example.

Soil nutrient status: fertile soil (with a high nitrogen content) favours an increase in the number of tillers.

Generally speaking, the shorter the sowing and the earlier the planting, the greater the number of tillers per plant.

The final number of stems depends on the number of tillers that survive to produce epis. The number of tillers and their survival rate are influenced by Autumn and winter weather conditions (for winter wheat). Cold weather slows

leaf and tillers induction.
The application of plant growth regulators generally serves to reduce apical dominance and increase the quantity and survival rate of tillers.
Nitrogen applications increase the size of the leaves and the number of flowers, as well as their chances of survival.

Chapter II

General points on growth modelling

I. Context and issues

Plants play an essential role in human survival and the balance of our ecosystem. Since the beginnings of agriculture, mankind has cultivated certain varieties of plant to meet its needs, in particular for food, clothing, heating or, more recently, to produce biofuels or extract molecules of therapeutic interest. In many cases, these plants require quantities of nutrients and water during their growth cycle if they are to develop properly (Brisson et al, 2002).

Fertilisation is a very important step in ensuring good crop growth and development, as well as optimum yields. Certain nutrient levels are very important for good crop growth, and optimising fertilisation strategies is an important issue in agronomy, which modelling-based approaches could help to address.

II. Agronomic modelling

Modelling is commonly used in agronomy to help researchers and scientists better understand the complex links between human actions, the pedoclimatic context and the responses of the agroecosystem. The models are also used for decision-making purposes, enabling a systemic analysis of the consequences of a change in crop management and an assessment of the risks associated with such changes.

III. The importance of modelling in agronomy

Modelling made its appearance in the field of agronomy 25 years ago with the work of de Wit (1978) on photosynthesis and respiration, and now occupies an important place. Taking advantage of the possibilities opened up by the development of information technology, modelling has become an essential tool for understanding the mechanisms involved in crop production and for inventing new techniques. In fact, simulation through crop models offers

on the one hand, the opportunity to extrapolate the knowledge gained from a small number of experiments to a wider range of conditions.

On the other hand, it allows the effects of different factors on the performance of the system under study to be quantified simultaneously (Boote et al. 1996). It also offers the possibility of exploring a wider range of situations in a shorter time span (Semenov et al. 2009). In addition, models are tools that give access to a variety of indicators that are difficult to access by experiment, such as solute or gaseous compound fluxes. They can also be used to understand very long-term changes in cropping systems.

IV. Basic notions of modelling

A model is "a simplified formulation that imitates real-world phenomena in such a way that it enables us to understand complex situations and make predictions. In their simplest form, models can be verbal or graphic, i.e. made up of concise statements or pictorial representations.

"(Odum, 1975) in (Gate, 1995).

A model can be expressed in the following simple mathematical form (Minet &Tychon, n.d.):

Equation 1

F = M (x. p)

Where Y represents the variable(s) explained by the model (the output variable); X the model input(s) and p the model parameters (Minet &Tychon, n.d.). The input variables are the variables observed or measured in all situations where the model is applied. For ecophysiological models that can describe crop growth and development, these may be meteorological data, characteristics related to cropping practices (sowing date, fertiliser application dates and doses), etc. (Dumont et al. 2012). the model parameters are values that are constant for all the situations studied. To return to the eco physiological models, the parameters of these models can be the water content at field capacity and at the wilting point. For each soil horizon, these values are constants used to calculate the soil's water reserve (Dumont et al. 2012). It is therefore important to find the best possible parameters in order to obtain output values that correspond to the observations. For this reason, the calibration stage is an important step in model creation (Minet &Tychon, n.d.).

4-1-Types of models

There are different types of model. Among these, two types have been selected:

4-1-1- Statistical models

Constructed from statistical laws, these models are based on relationships such as regression on one or more explanatory variables. They are simple to use and enable the most important explanatory variables to be identified quickly (Gate, 1995).

4-1-2- Mechanical models

These models are more complex because they integrate the main processes involved in the system under study (El Hassani &Persoons, 1994). For example, for models used in ecophysiology, the aim is to reproduce the reactions of the plant or crop. They are based on known physiological laws and functions. Different input data are used to feed these models: data of internal origin concerning plant properties (genetics, etc.) and data of external origin (meteorological, pedological data, etc.) (Gate, 1995; Rauff & Bello, 2015). The main difficulty with these models is their validation (El Hassani &Persoons,

1994).

4-2- Creating models

The first step in creating a model is to acquire knowledge about the phenomenon we wish to study. This will provide a basis for making different choices, particularly regarding the explanatory factors. Next, it is important to define the domain of use of the model, which will enable the experimental design to be defined (Gate, 1995).

The final stage involves the actual construction of the model, and therefore the choice of mathematical or statistical tools for processing the data. This choice will depend on the type of data to be used and the purpose of the model (Gate, 1995).

4-3 Model calibration

Calibrating a model involves adjusting its parameters in order to improve the prediction of output variables. The aim of calibration is therefore to improve the model. This operation is carried out under specific climatic or other conditions depending on the study area in which the model will be used. Once it has been produced, the calibrated model will have a predictive value that is more local than universal (Minet and Tychon, n.d.).

4-4-Validation of models

The validation phase is a key stage in model design. It will determine whether or not the model accounts for the phenomenon under study with sufficient precision, or whether all situations can be explained with the same level of precision (Gate, 1995).

The first step in validation may be to compare observed and simulated values using statistical tests. In addition, the analysis of residual errors between simulated and observed values is often considered in order to assess the performance of the model (Dumont et al, 2012).

The simplest method is to calculate the difference between the measured variable () and the variable estimated by the model (y):

Equation 2

$$= \gamma - \gamma'$$

The bias can then be calculated as follows:

Equation 3

$$Biais = \sqrt{\frac{1}{N}} + \overline{\sum_{i=0}^{n} D} = 1$$

In this equation, N represents the number of pairs of values (measurement - estimate). However, this method is insufficient to judge the quality of a model. This is why the square root of the root mean square error (Rmse, RootMean

Square Error) is used (Dumont et al, 2012).

Equation 4

$$RMSE = \sqrt{\frac{1}{N} \sum_{i=0}^{n} (\gamma - \gamma)}$$

It is easy to interpret the value obtained by the RMSE because it is expressed in the same units as the observed values. However, care must be taken with the meaning of this value, as greater weight is given to the largest errors by squaring the bias. The ideal value for the mean square error is 0 (Dumont et al., 2012).

V. The simple lincar regression model

Linear regression is a modelling method used to establish a linear relationship between a continuous variable known as the "explained" or dependent variable and a set of other continuous variables known as the "explanatory" or independent variables.

More specifically, it proposes an explanatory model that predicts the dependent variable as a function of the independent variables.

This module is devoted to the study of simple linear regression to model the predictive relationship between the dependent variable and a single independent variable. This modelling is used to develop the basic concepts of multivariate regression.

Regression can be used to replace a variable that is difficult to observe with another variable that is relatively simple to measure (Louis , 2016).

5-1-Regression line

To begin with, let's use the simplest model to express the relationship between X and Y. The relationship between the two characteristics is represented graphically by a regression line **y = a.x + b** .

In addition, the line passing "most centrally" through the point cloud corresponds to the line for which the sum of the squares of the vertical deviations of its points is minimal. In other words, our model will take into account the dispersion of the point cloud to calculate the estimated coefficients a and b of the equation of the regression line. In mathematics, this method **is called the least squares line** (Guillaume, 2020).

For this method, we obtain the following coefficients:

$$a=\frac{Cov(x,y)}{V(x)} \qquad \bar{y}=\frac{1}{n}\sum_{i=1}^{n} Y_i$$

$$b= \bar{y}-a\bar{x} \qquad \bar{x}=\frac{1}{n}\sum_{i=1}^{n} X_i$$

5-2-Explained variance and residual variance

As we saw earlier, the least squares method is used to obtain the line which minimises the sum of the squares of the differences between the predicted and observed Y values. This means that the regression can be written as Y = a.x + b + E , where **E is the random variable of the** vertical **deviation** of each point from the predicted value (green lines on the previous graph).

First, we can show that the random variables a.**x** + **b** and E are independent. We then obtain two variance formulae, the explained variance **Var(a.x+b)** and the residual variance **Var(E)**. Finally, these variances can then be taken into account in hypothesis tests of the correlation coefficient (Guillaume, 2020).

5-3- Regression quality

Finally, linear regression can be used to assign a quality indicator called the **coefficient of determination R^2** . This indicator is well known to MS Excel users, since it can be obtained at the same time as the plot and the equation of the regression line.

To obtain this, we need to compare the differences between the scatter plot and the predictive model obtained. From the trace of the regression line (y_i), the centre of gravity (y) and the point cloud (y_i), we calculate the following **analysis of variance table**:

Table . 2Analysis of variance in linear regression

Source of variation	Definitio n	Formul e
SCE	Sum of Explained Squares	$SCE = \sum(\hat{y}_i-\bar{y})^2$
SCR	Sum of Residual Squares	$SCR = \sum(y_i-\hat{y}_i)^2$
SCT	Sum of Total Squares	$SCT = SCE + SCR$

The calculation of SCE and SCR is partly reminiscent of the calculation of a variance, because we have the sum of the deviations from a squared reference value.

$$R2 = \frac{SCE}{SCT}= 1-\frac{SCR}{SCT}$$

The value of R^2 is then taken into account in our reflection. $R^2 = [0;1]$.

- **If R2 = 0** then the model explains nothing, the variables X and Y are not

linearly correlated.
- **If R2 = 1**, then the points are aligned on the line, and the linear relationship explains all the variation.
- Between these two values, it's up to you to judge whether the linear relationship is valid or not.

In conclusion, this coefficient of determination also allows us to judge the correlation between our two random variables X and Y. However, beware of the **stork effect**! Correlation is not causality, and statistics stop where reason takes precedence. You will therefore need to delve into the scientific literature to conclude whether your linear regression is biologically relevant...(Guillaume, 2020)

Chapter III

Materials and methods

I. Location of the experiment

This work was carried out at Jardin de I'universite de 20aout 1955 a Skikda, faculte des sciences, departement d'agronomie hors sol.

II.Equipment used

2-1-Laboratory equipment

we used the following tools

- Balancenumerique
- Nitrogen
- Petri dish
- spatula

Digital balance Nitrogen petri dish   spatula

2-2-Growing equipment

we used the following tools

- Pot
- Soil
- Shovel

- Axe

2-3-Plant material used

The study focused on four varieties of durum wheat (Triticumdurum), both local and imported: Amar 06, Antalis, Simeto and oued el bared. Supplied by the subsidiary, Constantine cerealiere, the industrial and commercial complex of El Harrouch (Agrodiv).

The origin of these varieties **is shown in table 04** below.

Table 03 The pedigree and origin of the varieties studied

Name	Pedigree	Origin
Simeto	Capeiti x valvona	Itali
Antalis	S/BitSCD26406	Itali

| Amar 06 | ID94.0920-C-OAP.7AP | Algeria |
| Oued el bared | GTA dUR /OFANTO-DZ- ITGC-SET-008-2004/2005-15-3S-0S | Algeria |

III. Pedological characteristics

The soil is a support for vegetation and crops, and the physical and chemical properties of the soil have a considerable influence on yields and crop health.

Table No. 04 shows the pedological characteristics of the soil used in our study.

Elements	Means
Ph	6,6
Conductivity (ms/cm)	/
C/N	/
Carbonates (%)	/
Potassium (ppm)	0,5
Magnesium (meq/100g)	4,0
Calcium (meq/100g	8,6
Sodium (meq/100g	0,7
Clay (%)	320
Sand (%)	220
Silt (%)	210

IV. Experimental method

The trial in plastic pots with a diameter of 14 cm and a height of 16 cm (figure08) was set up in the garden of the University of Skikda (Complexe Massoud Boukadoum). 4 genotypes were used, with an average of 8 repetitions for each genotype:

4 genotypes x 8 iterations = 32 distributed experimental units.

According to table 05 breakdown of experimental units **(Experimental set-up)**

\ Varieties Repetitions^	Variety s			
	G1 (1)	G2(1)	G3(1)	G4 (1)
	G1 (2)	G2(2)	G3(2)	G4 (2)
Repetitions	G1(3)	G2(3)	G3 (3)	G4 (3)
	G1(4)	G2(4)	G3	G4 (4)

		(4)	
G1(5)	G2(5)	G3(5)	G4(5)
G1(6)	G2(6)	G3(6)	G4(6)
G1(7)	G2(6)	G3	G4

Figure 08. Soil cultivation of the varieties studied.

4-1-Imbibition and germination

This work was carried out in the greenhouse and soil chemistry laboratory of the University of 20 August 1955 in Skikda, Faculty of Science, Department of Agronomy. The wheat seeds (*Triticumdurum*) were placed in distilled water for 24 hours for imbibition on 19/12/2022.

The seeds were then germinated in Petri dishes covered with two layers of tissue paper moistened with water distilled at laboratory temperature, where we exposed one group of plants to light and placed the second group in total darkness (Figure.09).

We follow the seeds as long as we keep them moist until the coleoptile emerges and the first leaf, so that they can be planted in the soil.

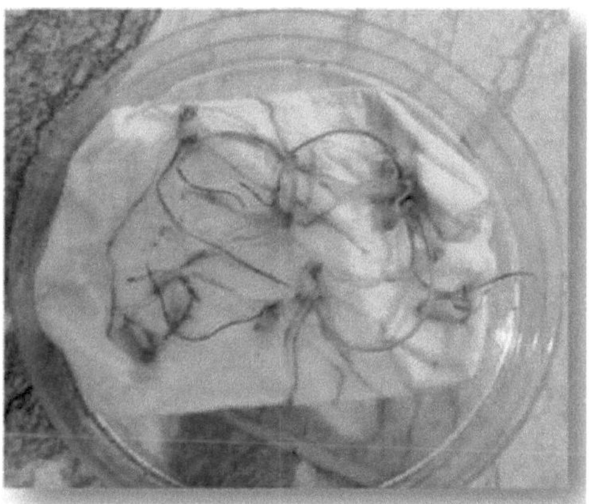

Figure 09: Seed germination test.

4-2-Planting

Sowing was carried out on 25/12/2022 at a density of 4 grains (germinated seeds) per pot, at a depth of approximately 1.5 cm by hand.

Figure 10planting seeds

4-3-Watering

The plants are regularly watered twice a week with
eaunormale as required until the end of the biological cycle.
The experimental conditions were carried out under natural conditions and were standardised in the experiments in terms of growing medium (soil) and climatic conditions (quantity of light, temperature and watering).

4-4-Technical directories

The trial is maintained on a regular basis by manual weeding and by supplementing with an Addition of nitrogen to four of the eight pots of each variety at the surface of the pots at the tillering stage (after the 4th leaf has been emitted) on 21/03/2023. Adding 4g per pot. Because it corrects the lack of nutrients in the soil.

4-5-Harvesting The harvest is carried out manually at the end of... for all varieties.

Figure . 11 Manual harvesting of varieties **Figure 12** Manual harvesting of varieties with treatmentswithout treatments

V. Nitrogen fertilisation

We added nitrogen to four of the eight pots in each replicate at the surface of the pots at the tillering stage (after emission of the 4th leaf) on 21/03/2023.

where we used this relationship to calculate the added nitrogen dosage.

1h >2000000g

S > X

Applying the relationship, we see that in the space of each pot 200.96 cm we put Dosage 4.01g of nitrogen and put it in each pot.

VI. Measured parameters

7-1-Characters

morphology 7-1-1The number of tillers

Was determined by selecting eight pots in each variety four pots before treatment and four pots after treatment (nitrogen fertilisation) and comparing the results.

7-1-2- Plant height

Plant lengths were measured from the beginning of the stem to the end of the beards during the ripening phase.

7-2-Performance components

7-2-1-Number of ears per pot (NE/pot)

The number of ears in each pot was counted directly.

7-2-2-Epis languor

The epis languor has been calculated from the epis pass expremite to the final epis summit.

7-2-3-The number of grains/epis

The number of grains per ear was counted directly.

7-2-4- The weight of 1000 grains

The 1000grain point was estimated on the basis of available grain.

7-2-5-Nitrogen dosage

Was determined by selecting four repetitions of eight for each variety.

VII.Statistical analysis methods

Statistical analysis of the data obtained was carried out using Excel stat ANOVA software for analysis of variance.

I. Morphological study of tillering

According to our results, this stage begins with the elongation of the stem and the emergence of the first leaf until the emergence of the seventh (07) leaf. In the course of following the plants, we recorded the dates in the following table 06

Table 06. Leaf emission dates.

Variety N^0 leaves	Simeto	Oued el bared	Amar 06	Antalis
1st sheet	7 days	8 days	7 days	8 days
2ndleaf	15 days	18 days	17 day	17 day
3rd leaf	29 day	31 days	30 days	31 days
4th leaf	36 days	39 day	38 day	40 days
5th leaf	45 days	46 days	45 days	47 days
6th leaf	55 days	57 days	56 days	57 days
7th leaf	62 days	64 days	63 days	63 days

We have entered the dates on which the tillers of the varieties **studied** were issued **in table 8 below.**

Table 07 . Tiller emission dates

Varieties No. of tillers	Simeto	Oued el bared	Amar 06	antalis
1 heel	36 days	38 day	37 days	38 day
2 tillers	43 days	45 days	44 days	45 days
3 tillers	50 days	/		//

We monitored the appearance of tillers (from the early tillering stage to the late tillering stage).

According to our results, the appearance of the 1st tillers begins when the seventh leaf is emitted. The formation of the tillers is accompanied by the formation of adventitious roots in order to meet the nutritional requirements of these new organs.

II. Effect of nitrogen doses on the components of yield 2-1- Effect of nitrogen dose on the number of tillers

With treatment (nitrogen fertilisation)

The study of the revolution of the number of tillers in the four varieties after the treatment (added nitrogen) figure 13, this result showed that the components respond positively to nitrogen fertilization so that the recorded results give that the between 12 and 7 tillers, in exchange the lowest values were recorded for the variety Antalis and Oued el bared between 10 and 5 tillers. The results show that nitrogen increases the number of tillers, so nitrogen fertilisation improves tillering.

Number of tillers for varieties treated with nitrogen

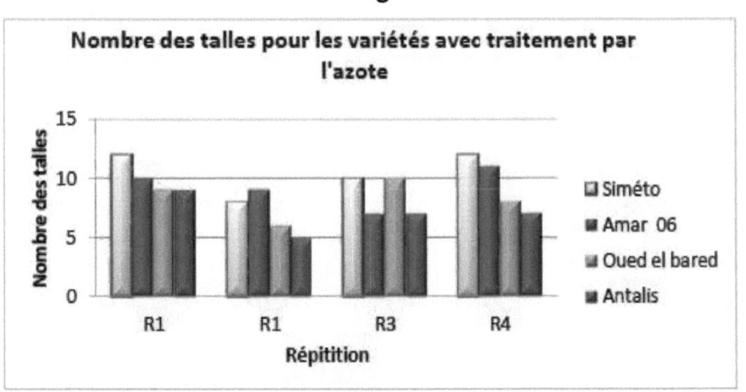

Figure N°13 . Number of tillers for varieties treated with nitrogen

Table 08. Number of tillers for varieties treated with nitrogen

N° of tillers variety	R1	R2	R3	R4	$\square$	Variance	S
Simeto	12	8	10	12	10, 5	3,66	1,9 1
Amar 06	10	9	7	11	9,2 5	2,91	1,7 0
Oued el bared	9	6	10	8	8,2 5	2,97	1,7 1
Antalis	9	5	7	7	7	2,66	1,6 3

without treatment

The study of the evolution of the number of tillers in the four varieties without nitrogen treatment (Figure N°14) showed that the number weighed between 6 and 1 tillers where the highest values were recorded in the variety simeto with 6 tillers and amar 06 5 tillers while the lowest values were recorded in the varieties oued el bared and antalis between 4 and 1 tillers.

Number of tillers for varieties not treated with nitrogen

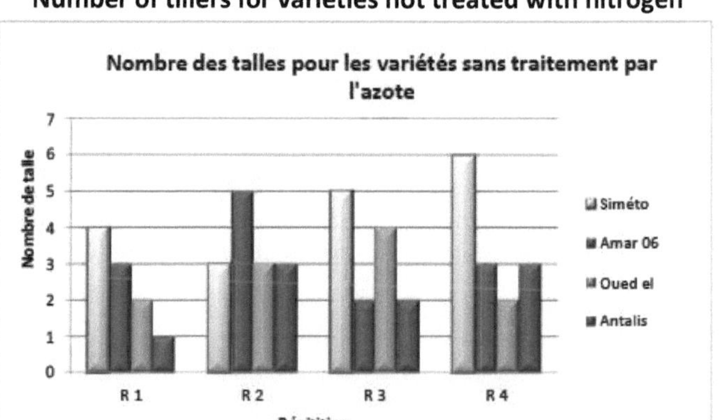

Figure N°14 . Number of tillers for varieties without nitrogen treatment

Table 09. Number of tillers for varieties without nitrogen treatment

N° of tillers variety	R1	R2	R3	R4	X	Varianc e	S
Simeto	4	3	5	6	4,5	1,66	1,2 8
Amar 06	3	5	2	3	3,2 5	1,58	1,2 5
Oued el bared	2	3	4	2	2,7 5	0,91	0,9 5
Antalis	1	3	2	3	2,2 5	0,91	0,9 5

2-2- Effect of nitrogen dose on plant height

Without treatment

The study of the revolution of the height of the plant in the four varieties without treatment by nitrogen (Figure N°15), showed that the height between 55 cm and 49cm where the highest values were recorded in the variety Simeto with 55cm and amar06 with 50cm whereas in the variety amar06 with 50cm.

the lowest values were recorded for the oued el bared and Aantalis varieties between 44 cm and 48 cm.

Plant height for varieties not treated with nitrogen

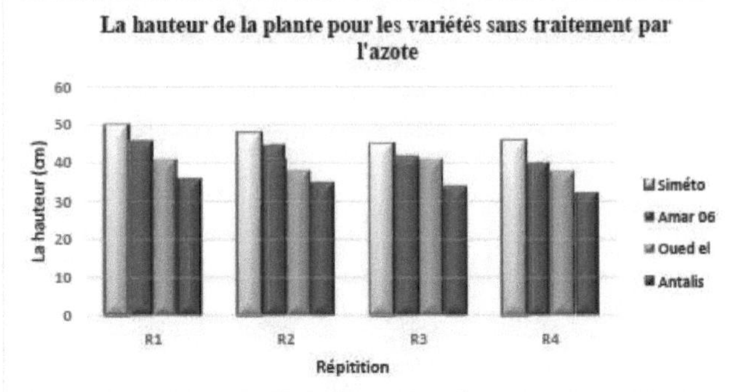

Figure N°15 . Plant height for varieties without nitrogen treatment

Table 10. Plant height for varieties without nitrogen treatment

No. of tillers variety	R1	R2	R3	R4	X	Variance	S
Simeto	50	46	41	36	43,2 5	36,91	6,0 7
Amar 06	48	45	38	36	41,5	36,33	6,0 2
Oued el bared	45	42	41	34	40,5	21,66	4,6 5
Antalis	46	40	38	32	39	33,33	5,7 7

With treatment (nitrogen fertilisation)

The study of the evolution of the plant height in the four varieties after the treatment (Figure N°16), this result showed that the compasants respond positively to the nitrogenous fertilization so that the recorded results give that the plant height between 36 and 63 cm, the highest values were recorded in the varieties Simeto and Amar 06 between 50 and 63 cm, in exchange the lowest values were recorded in the variety Antalisbared between 36 and 45 cm. The results showed that nitrogen fertilisation improved plant height. Plant height for varieties treated

with nitrogen

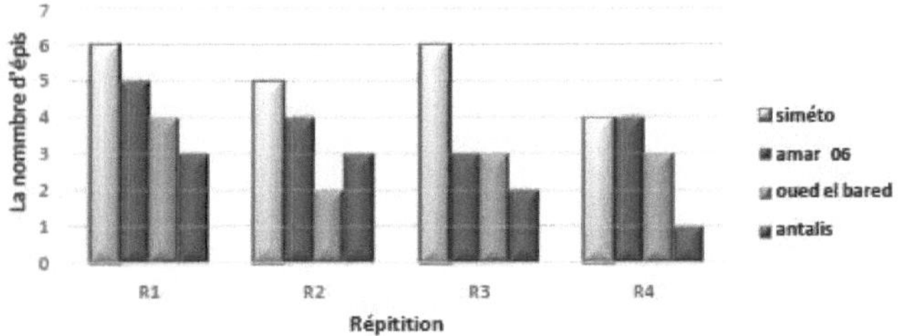

Figure N°16 . Plant height for varieties without nitrogen treatment

Table 11. Plant height for varieties without nitrogen treatment

No. of tillers variety	R 1	R 2	R 3	R4	X	Varianc e	S
Simeto	63	60	59	60	60,5	3	1,73
Amar 06	59	58	55	50	55,5	16,33	4,04
Oued el bared	45	42	41	40	42	4,66	2,15
Antalis	41	40	38	36	38,7 2	5,70	2,38

2-3- Effect of nitrogen dose on the number of spikes/pot

J **No treatment**

The study of the revolution of the number of ears in the four varieties without treatment by nitrogen (Figure N°17), showed that the number of ears per pot between 6 and 1 ear or the highest values were recorded in the variety Simeto with 6 tillers and Amar 06 with 5 tillers while the lowest values were recorded in the varieties Oued el bared and Antalis between 4 and 1 ear.

The number of epis for varieties without nitrogen treatment

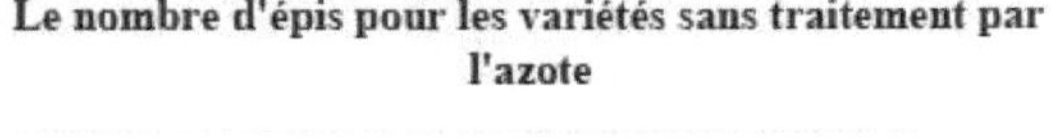

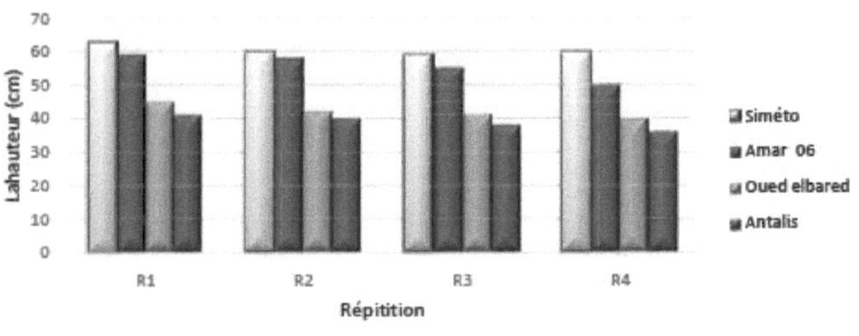

Figure N°17 . The number of udders for varieties not treated with nitrogen

Table 12 . number of udders for varieties not treated with nitrogen

No. of tillers variety	R 1	R 2	R 3	R 4	X	Variance3	
Simeto	4	3	5	6	4,5	1,66	1,2 8
Amar 06	3	5	2	3	3,2 5	1,58	1,2 5
Oued el bared	2	3	4	2	2,7 5	0,91	0,9 5
Antalis	1	3	2	3	2,2 5	0,91	0,9 5

With treatment (nitrogen fertilisation)

The study of the revolution of the number of ears per pot in the four varieties after the treatment (Figure No. 18), this result showed that the compasants respond positively to nitrogen fertilization so that the results recorded that the number of grains between 12 and 5 ears, The highest values were recorded for the simeto and Amar 06 varieties between 12 and 9 ears, while the lowest values were recorded for the Antalis and Oued el bared varieties between 9 and 5 ears. The results showed that nitrogen fertilisation improved the number of ears per pot.

The number of plants per pot for variëtës treated with nitrogen

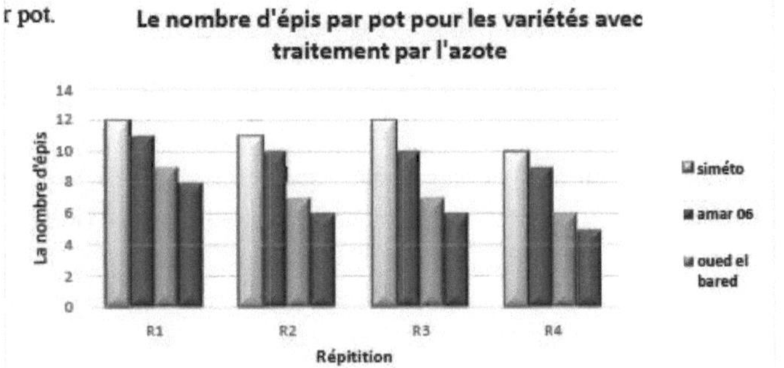

Figure N°18 . number of pis per pot for varieties without nitrogen treatment

Table 13: Number of plants per pot for varieties not treated with nitrogen

N°of tillers variety	R1	R 2	R3	R 4	X	Variance	S
Simeto	12	8	10	12	10, 5	3,66	1,9 1
Amar 06	10	9	7	11	9,2 5	2,91	1,7 0
Oued el bared	9	6	10	8	8,2 5	2,97	1,7 1
Antalis	9	5	7	7	7	2,66	1,6 3

2-4- Effect of nitrogen dose on the number of grains per ear

With treatment (nitrogen fertilisation)

The study of the change in the number of grains per spike in the four varieties after treatment (Figure No. 19) showed that the compasants responded positively to nitrogen fertilisation, so that the results recorded gave the number of grains between 44 and 30 grains, the highest values were recorded in the Simeto and Amar 06 varieties between 44 and 42 grains, while the lowest values were recorded in the Antalis and oued el bared varieties between 36 and 30 grains.The results showed that nitrogen fertilisation improved the number of grains per ear.

Number of grains per ear for varieties treated with nitrogen

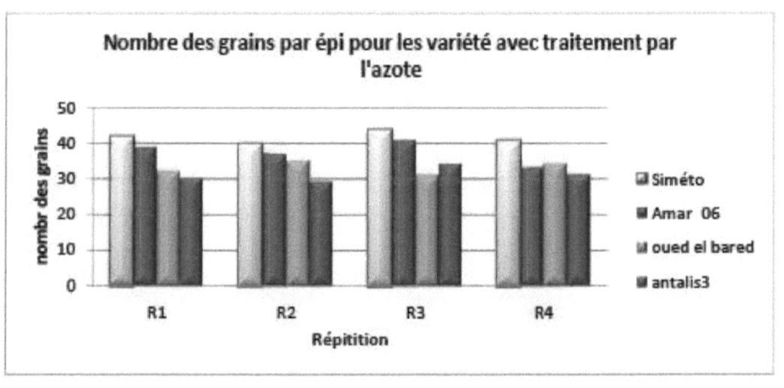

Figure N°19 . number of grains per ear for varieties treated with nitrogen

Table 14: Number of grains per ear for varieties treated with nitrogen

No. of grains variety	R1	R2	R3	R4	X	Variance	S
Simeto	42	40	44	41	41,7 5	2,91	1,70
Amar 06	39	37	41	33	37,5	11,66	3,41
Oued el bared	32	35	31	34	33	3,33	1,82
Antalis	30	29	34	31	31	4,66	2,15

Without treatment

The study of the evolution of the number of grains in the four varieties without nitrogen treatment (Figure N°20) showed that the number of grains between 30 and 20 grains, the highest values were recorded in the varieties Simeto and Amar06 between 30 and 24 grains compared to the other two varieties Oued el bared and Antalis of (23 and 19) respectively.

Number of grains per epis for varieties without

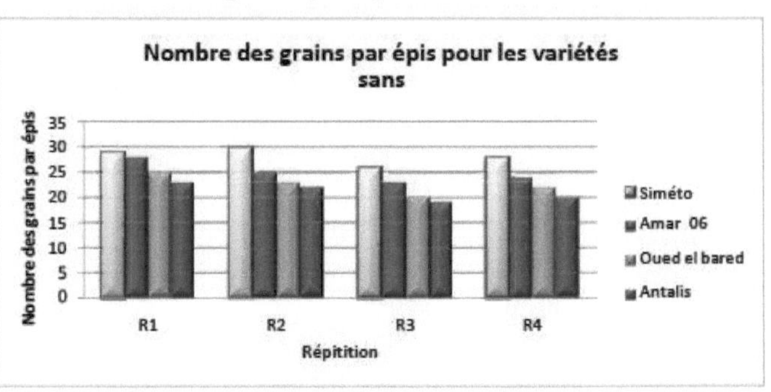

Figure N°20 . Number of grains per ear for varieties without nitrogen treatment

Table 15. Number of grains per ear for varieties without nitrogen treatment

No. of grains variety	R1	R2	R3	R4	X	Variance	S
Simeto	29	30	26	28	28,2 5	2,91	1,70
Amar 06	28	25	23	24	25	4,66	2,15
Oued el bared	25	23	20	22	22,5	4,33	2,08
Antalis	23	22	19	20	24	15,33	3,9

2-5- Effect of nitrogen dose on earlength
With treatment (nitrogen fertilisation)

The study of the revolution of the length of ears in the four varieties after the treatment (Figure 21), this result showed that the composites respond positively to nitrogen fertilization so that the results recorded give that the length of ears between 8 and 6 cm, The highest values were recorded in the Simeto and Amar 06 varieties between 8 and 7.1 cm, while the lowest values were recorded in the Antalis and oued el bared varieties between 6.15 and 6 cm.The results showed that nitrogen fertilisation improved ear length.

Length of epis for varieties with nitrogen treatment

Figure N°21. Epis lag for varieties with nitrogen treatment
nitrogen

Table 16. Length of epis for varieties with nitrogen treatment

No. of grains variety	R1	R2	R3	R4	X	Variance	S
Simeto	7,5	7	8	7,0 8	7,39	1,39	1,17
Amar 06	7	6 ,68	7,1	6,2 5	6,75	0,14	0,37
Oued el bared	6,1 9	6,50	6,15	6,3 3	6,29	0,025	0,15
Antalis	6	6,1	6,25	6,0 6	6,10	0,01	0,1

Without treatment

A study of the evolution of earlength in the four varieties without nitrogen treatment (Figure No. 22) showed that earlength ranged from 6 to 5.20 grains, with the highest values recorded in the Simeto and Amar 06 varieties, at between 6 and 5.88 cm, compared with the other two varieties, Oued el bared and Antalis, at between 23 and 19 cm respectively.

Epis lag for varieties without nitrogen treatment

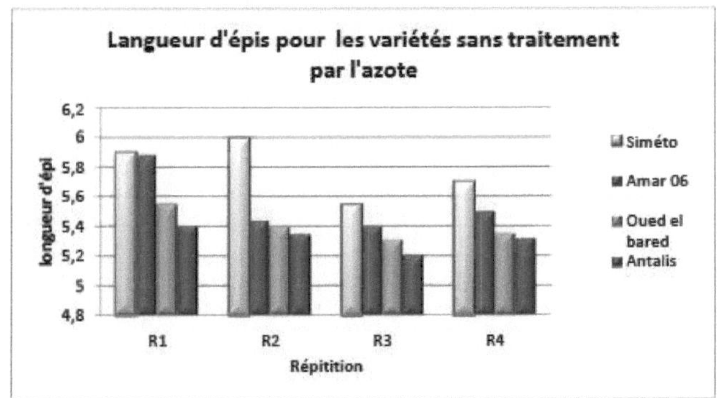

Figure N°22 . Epis lag for varieties without nitrogen treatment
nitrogen

Table 17. Length of epis for varieties without nitrogen treatment

Grain no. Variety	R1	R2	R3	R4	X	Variance	S
Simeto	5,9	6	5,5 5	5,7	5,78	0,04	0,2
Amar 06	5,8 8	5,4 4	5,4	5,5	5,55	0,04	0,2
Oued el bared	5,5 5	5,3 9	5,3	5,3 5	5,39	8,53	2,9
Antalis	5,4 0	5,3 4	5,2 0	5,3 1	5,3	7,23	2,6 8

2-6- Effect on the weight of 1000 grains

Without treatment

Figure No. 23 shows that one thousand untreated grains weighed between 31.19g and 15.9g. Where the highest values were recorded in individuals Simeto as the highest value and Amar 06 as the lowest.
the lowest values were recorded in Antalis and Oued el bared.

Weight of 1000 grains without nitrogen treatment

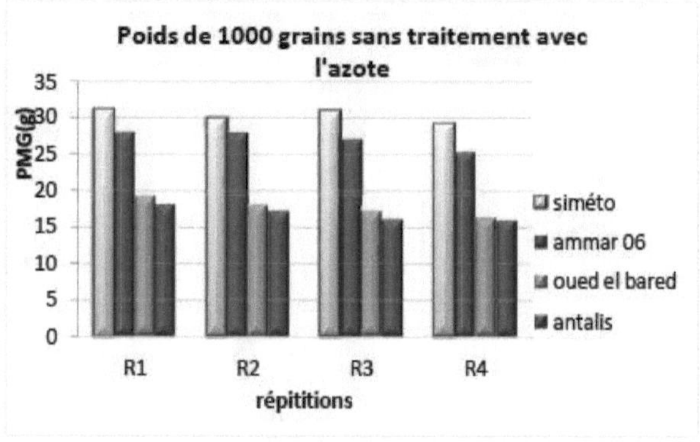

Figure N°23 . Weight of 1000 grains for varieties without nitrogen treatment

Table 18. Weight of 1000 grains for varieties without nitrogen treatment

Grain no. Varierex.	R1	R2	R3	R4	X	Varianc e	S
Simeto	31,1 9	30	31,0 3	29,2 5	30,3 6	0,82	0,9 0
Ammar 06	28	27,9	27	25,3	27,0 5	1,92	1,3 8
Oued el bared	19,3 1	18	17,2 2	16,3 7	17,7 2	1,55	1,2 4
Antalis	18,1	17,1 6	16,0 5	15,0 9	16,8 0	1,06	1,0 2

Tables 10 and 10Average analysis of variance **of morphological** characteristics **and yield components with and without nitrogen treatment**.

With treatment

A study of the evolution of the weight of 1000 seeds treated with nitrogen (Figure 24) showed that this component responds positively to nitrogen fertilisation.

In fact, a single nitrogen application at tillering produced the lowest PMG of 44.13g, while the highest PMG was 55.5g in the Simeto variety, for which

nitrogen fertilisation improved the PMG of durum wheat (Mandic et al; 2015).
Weight of 1000 grains treated with nitrogen

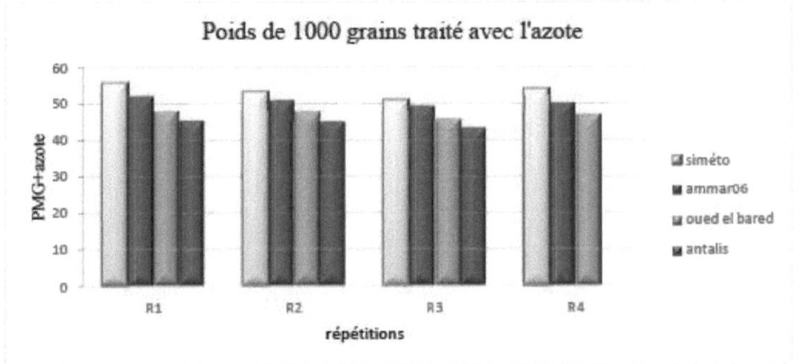

Figure N°24 . Weight of 1000 grains treated with nitrogen for varieties
without nitrogen treatment

Table 19. Weight of 1000 grains treated with nitrogen for varieties without
nitrogen treatment

Grain no. Variety	R1	R2	R3	R4	X	Variance	S
Simeto	55, 8	53, 3	51,1 5	54,0 1	53,5 6	3,69	1,9 2
Ammar06	52, 1	51	49,3	50,1 2	50,6 3	1,43	1,1 9
Oued el bared	48	47, 7	45,5 2	46,7 8	47	1,24	1,1 1
Antalis	45, 3	44, 8	43,0 9	44,1 3	44,3 3	0,92	0,9 5

2-7- Nitrogen fertilisation

We add nitrogen to four of the eight pots in each repetition at the surface of the
pots at the tillering stage (after the 4eme leaf has been removed on 21/03/2023)
where we used this relationship to calculate the dose of nitrogen added.

1h >2000000g

S > X

1x108cm2 >2000000g

200.96cm x^2

X = 401920000 -1x108

X = 4,01

We added 4g of nitrogen to each pot.

II. Analysis of variance of results

Table 20. Analysis of variance for number of tillers

Source de Variance L	DD	SC E	CM	F-obs	F-théo
Entre Groupes	3	26,5	8,83	2,90	3,49
A l'intérieur des groups	12	36,5	3,04		
Totale	15	63			

The results of the analysis of variance (table 19) show that there is a significant difference for this morphological characteristic because

F observed < F theoretical

Table 21. Analysis of variance for plant height

Source ofVariance L	DD	SCE	CM	Fobs	F- theo
Between Groups	7	2209,5	315,64	53,34	2,42
Within groups	24	142	5,91		
Total	31	2351,5			

The results of the analysis of variance (table .20) show that there is a significant difference for this morphological characteristic because

F observed > F theoretical

Table 22. Analysis of variance of number of ears/pot

Source of Variance L	DD	SCE	CM	F-obs	F theo
Between Groups	318,345	,	437,002	,42	
Within groups	24	29,	51,22		
Total	31	347,87			

The results of the analysis of variance (table 21) show that there is a significant difference for this morphological characteristic because

F observed > F theoretical

Table 23. Analysis of variance for La langueur d'epis

Source of Variance	DD L	SCE	CM	F-obs	F- theo
Between Groups	715,5	82,26	21,	192,42	
Within groups	24	2,560	,10		
Total	31	18,42			

The results of the analysis of variance (table 22) show that there is a significant difference for this morphological characteristic because

F observed > F theoretical

Table 24. Analysis of variance in number of grains/ear

Source of Variance	DD L	SCE	CM	F-obs	F- theo
Between Groups	7	1410	201,42	41,14	7,51
Inside groups	24	117,5	4,89		
Total	31	1527,5			

The results of the analysis of variance (table 23) show that there is a significant difference for this morphological characteristic because

F observed > F theoretical

Table 25 . analysis of variance of 1000 grain weights

Source ofVariance	DDL	SCE	CM	F-obs	F- theo
Between Groups	7	6140,20	877,17	541,11	8,04
Within groups	24	38,90	1,62		
Total	31	6179,10			

The results of the analysis of variance (table 24) show that there is a significant difference for this morphological characteristic because :

F observe > F theor

II. Mathematical and numerical study of the model <u>For the linear model</u>

Qs = NTM.T + N

Qs = The quantity of biomass

T = time (per day)

NTM = development of the number of tillers

N = influencing substance (nitrogen)

This model develops linearly

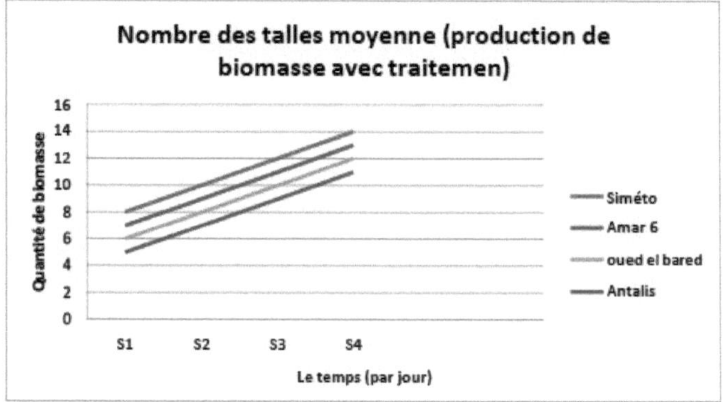

Figure . 25 Dynamic system of plant growth (Wallach et al, 2005).

<u>**By comparison**</u>

Qs = NTM.T

Qs = The quantity of biomass

T = time (per day)

NTM= growth in number of tillersOur model is linearly variable

Average number of tillers (biomass production with treatment)

Figure.25 Number of average tillers (biomass production with treatment)

Number of average tillers (biomass production without treatment)

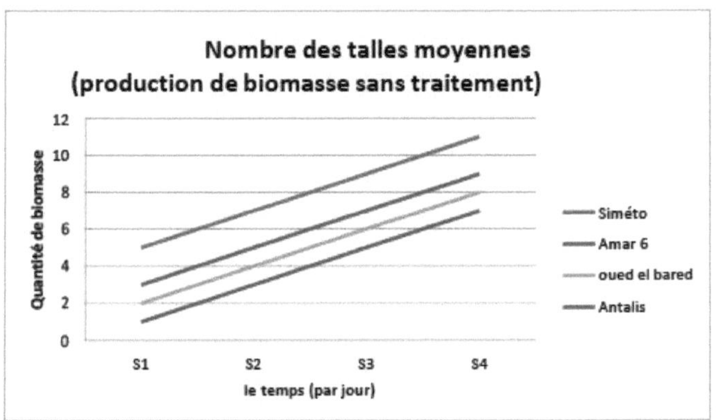

Figure . 26 Number of average tillers (biomass production without treatment)

General conclusion

The study was carried out on four varieties of durum wheat (*Triticumdurum*) most commonly used in Skikda: Simeto, Ammar 06, Oued el bared and Antalis (local and imported), with the aim of characterising the tillering of durum wheat (*Triticumdurum*) and its modelling under the effect of nitrogen.

Our main crop levels: eight replicates for each variety, including four untreated replicates and five treated replicates (nitrogen fertilisation).

We dealt with the following points:

- Morphological study of tillering.
- The effect of nitrogen dose on the number of tillers.
- The effect of nitrogen fertiliser splitting on plant height.
- Characterisation and modelling of tillering in durum wheat (*Triticumdurum*)

The effect of nitrogen fertiliser fractionation on production quality or yield components (number of spikes/m^2 , spike length in cm, number of grains per 1000 grain weight, nitrogen dose).

We have found that nitrogen fertilisation is a key element in increasing the number of tillers at the tillering stage using the **linear model**.

The ability to produce tillers is also greater in the Simeto and Amar 06 varieties than in the other Oued el Bared and Antalis varieties.

Our model can be made available to farmers producing durum wheat and our experiment on modelling the heel at the tillering stage will be a starting point to help our farmers grow durum wheat (*Triticumdurum)* and improve yields.

Finally, we recommend that farmers :

- Use dwarf seedling varieties for lodging.
- To optimise nitrogen fertilisation, you need to use modelling.

The Simeto variety is perfect for Skikda's climate.

References

- **Abbassenne, F., Bouzerzour, H., &Hachemi, L. (1997).** Phenology and production of durum wheat (*Triticumdurum* Desf.) in semi-arid zones. *Ann. Agron. INA*, El Harrach, 18 :24-36
Algerie : Universite Djilali Bounaama de Khemis Miliana, 2017, p 104.
Available on Algerie and ICARDA: 176 p
- **Acevedo, E., Silva, P., & Silva, H. (2002).** Wheat growth and physiology, *In:* Curtis, B. C., RajaramS., and Macpherson, G. H., Bread wheat. Improvement andProduction, Eds. *Food and AgricultureOrganization*, Rome, 30: 34 - 70.
- **-Alismail W et al, (2017).** Influence of sowing density on wheat durum production in the semi-arid zone of Haut Cheliff. These de mastere. Univ de Khemis- Miliana.51.
- **Baldy, C. (1984).** Efficient use of water by vegetation in Mediterranean climates. Bull. Soc.
- Belaid D., (1987). **Study of economic fertilization. Option mediterraneenneCIHEAM. 7 p**
- **Belaid, D. (1987).** Etude de la fertilisation azotee et phosphatee d'une variete de ble dur (Hedba3) enconditions de deficit hydrique, Memoire de magister. INA - El Harrach, Algiers, 108p
- **Bonjean, A. (2001).** History of cereal cultivation and in particular that of bletendre (Triticum
- **Bos H.J. , Neuteboom J.H. 1998-** Morphological analysis of leaf and tiller number dynamics of wheat (Triticum aestivum L.) : responses to temperature and light intensity . Annals of Botany . 81 : 131-139 .
-**Boukensous W**. Study of the efficacy of some fungicides on the control of foliar diseases of wheat and the impact of the treatment on the
crop development and yield [On line].
- **Boulal H., El Mourid M., Rezgui S., Zeghouane O. 2007**:
Guide pratique de la conduite des cereales d'automne (bles et orge) dans le Maghreb (Algerie, Maroc, Tunisie). Published by ITGC, INRA Algerie and ICARDA: 176 p.
- **Brink, M., &Belay, G. (2006).** Ressources végétales de l "Afrique tropicale 1. Cereals and pulses. PROTA Foundation, Wageningen, The Netherlands/BackhysPublishers, Leiden, The Netherlands/CTA, Wageningen, The Netherlands, 328pp.
- **Casnin C., Jean-Francois M. and Levesque H. (2013**). Le ble, une plante modèle pour etudier labiologie végétale au lycee (teachers-associates at Ife-ENS Lyon)

-**Catell, F., 2006-** Hydric and physiological functioning of the plant. In: TiercelinJ.R.et Vidal cereales d'automne (bles et orge) dans le Maghreb (Algerie,Maroc, Tunisie). Edition: ITGC, INRA

-   **Cherfia, R. (2010).** Etude de la variabilite morpho-physiologique et moleculaire d'une collectionde ble dur algerien (**Triticumdurum**). En vue de Γ obtention du diplôme de Magistere enBiotechnologies végétales Universite Mentouri, Constantine

-   **Clement J.M. 1981**: Dictionnaire Larousse Agricole. Librairie Larousse. ISBN2-03- 514301-2: 1207p.

-   **Clerget, Y. (2011).** Biodiversity of cereals Origin and evolution. Montbeliard. 17p. of two Algerian hard wheat varieties (Bousseleme and Simeto) [Online]. Master's thesis.

-   **Diehl, R, 1975**. Agriculture generale. Edition J.B. Bailliere. 396 pages. and the impact of treatment on crop development and yield [Online].

-   **Dubcovsky, J., & Dvorak, J. (2007).** Genome plasticity a key factor in the success of polyploid wheat under domestication. Science, 316 (5833) :1862.

-   **Duthil, J., 1973** - La fertilisation phosphatee des sols calcaires. An Agro, INA VolVI n°2, pp.

-   **Emillie (2007).** Connaissance des aliments base alimentaire et nutritionnelles de la dietetique. Ed:tec et doc, la voisier, paris.

-   **Even L.T. , 1975-** Photosynthesize and the flag leaf and components of bordering grain development in wheat.Aust j , biol , Sci , 23 ; 245p .

-   **Feldman, M., & Sears, E. R. (1981).** The wild gene resources of wheat. Sci. Am,244 : 98-109.

-   **Gallais, A., &Bannerot, H. (1992).** Amelioration des espepees végétales cultivtives: objectifs et criteresde sélection. *INRA editions*. 759 pmanagement in Kentucky. The Univ. of Kentucky. http://www.uky. edu/Ag/GrainCrops/ID125 Section2.html (accessed 29 Nov. 2012).

-**Gate P, 1995**: Ecophysiologien du ble de la plante a la culture -Ed. DOC-la voisior I.T.C.F- France-pp 417.

Gate, P. H. (1995). Ecophysiologie du ble ; Technique et documentation : Lavoisier,Paris, 429 p

-**Girard M.C., Walter C., Remy J.C., Berthelin J. & Morel J.L.,**(2005).Sols et Environnement, Eds, Dunod, Paris, 816p.

-**Gouasmi R, Badaoui N.** Biochemical study of the influence of drying on nutritional value.

-**Guillaume, 2020.**les basesenbiostatistiques, lournos nature-2019-2022.

-**Grignac, P. (1965).** Contribution a l'etude de

TriticumdurumDesf. (Doctoral dissertation, Toulouse).

- **Hamadache, A. (2013).** Elements de phytotechnie generale : Grandes Culture- TomI : Le ble. 1ereedition. Mohamed Amrani. 49-69.
- **Hayden, B., (1990).** Nimrods, Piscators, Pluckers and Planters: The Emergence of Food Production. J. Anthrop. Archaeol, 9(1), 31
- **Henry, Y., & De Buyser, J. (2001).** L "origine des bles. In: Belin. Pour la science (Ed.). De la graine a la plante. Ed. Belin, Paris, pp. 6972.
- **Herbek, J., & Lee, C. (2009).** Growth and development. In: A comprehensive guideto wheat.
Karou, M., Haffid, R., Smith, D. N., & Samir, K. (1998). Roots and growth wateruse and water use.
- **Laumont, P., &Erroux, J. (1961).** Inventaire des bles durs rencontres et cultives en Algerie. Memoire de la societe d "histoire naturelle de PAfrique du Nord, 5 : 94p.le (01/03/2020).
- **Levy, A. A., & Feldman, M. (2002).** The impact of polyploidy on grass genome evolution. Plant Physiol, 130: 1587-1593
- **Longnecker, N., Kirby, E. J. M., & -e Robson, A. (1993).** Leaf emergence, tiller growth, and apicaldevelopment of nitrogen-deficient spring wheat.
Crop Sci, 33: 154- 160.efficiency of spring durum wheat under early- season drought. *Agronomy*, 18: 181-186.
- **Louis Houda, 2016.**PAF analyse de problem es de management
https://oraprdnt.uqtr.uquebec.ca/Gscdepot/paf1010/19 /M13.pdf
- **Mac Key, J. (2005).** Wheat: Its concept, evolution, and taxonomy. In: ConxitaMemoire de Master. Algerie : University 8 Mai 1945 Guelma, 2014, p 92.available on consulte
- **Michele M. Roger P. and Jean C. R. , 2006-** Biology and Multimedia UniversitéPierre et Marie Curie UFR de Biologie
Moeller C. Jochem B. , Evers and Greg R. , 2014- Canopy architectural and physiological characterization of near - isogenic wheat lines differing in the tiller inhibition gene tin . Frontiersin Plant Science : Plant Biophysics and Modeling.doi : 10.3389.
- **Mohamed, H. (2000)** Etude des systemes de production utilises en zone nord de Constantine casdu reseau d'amelioration du ble dur.
Moule , C. (1971) . Cereals . The Rustic House .
- **Oudjani W., 2009-** Diversite de 25 genotypes de ble dur

(TriticumdurumDesf.) : Etude des
production and adaptation characteristics. These de Magister.
Universite de Constantine. 111p.

-   **Robert, D., Gate, P., & Couvreur, F. (1993).** Les stades du ble.
Editions ITCF. 28 p.

-   **Sadouki M et *al*, (2018).** Etude de la variabilite morpho- physiologique du ble dur (*Triticumdurum.*) dans les conditions climatiques du Haut Cheliff. These de mastere.Univ de Khemis Miliana.

-   **Surget, A., &Barron, C. (2005).** Histology of wheat grain. Industries des cereales,(145), 3-7.

-**Soltner, D. (1990).** Alimentation des animaux domestiques. Tome 2, La pratique du rationnementdes bovins, ovins, porcins.

-**Wardlaw, I. F. (2002).** Interaction between drought and chronic high temperature during kernel fillingin wheat in a controlled environment. *Annals of Botany*, 90: 469-476.

Memoire de Master. Algerie : Universite 8 Mai 1945 Guelma, 2014, p 92. Available on consulted on (01/03/2020).

-**Ducreux G. 2002** - Introduction a la botanique (licence 1.2.3) .
Edition belin . Paris . p : 101-185 .

aestivum L.). Dossier de l'environnement de l'INRA, 21: 29-37

Buy your books fast and straightforward online - at one of world's fastest growing online book stores! Environmentally sound due to Print-on-Demand technologies.

Buy your books online at
www.morebooks.shop

Kaufen Sie Ihre Bücher schnell und unkompliziert online – auf einer der am schnellsten wachsenden Buchhandelsplattformen weltweit! Dank Print-On-Demand umwelt- und ressourcenschonend produzi ert.

Bücher schneller online kaufen
www.morebooks.shop